A Gentle Rain of Starlight

THE STORY OF ASTRONOMY ON MAUNA KEA

Michael J. West, Ph.D.

ISLAND HERITAGE™
PUBLISHING

Dedication

To Cheryl and Caden, who show me every day that the only thing

in the universe more precious than knowledge is love.

ISLAND HERITAGE™
P U B L I S H I N G
A DIVISION OF THE MADDEN CORPORATION

94-411 KŌʻAKI STREET, WAIPAHU, HAWAIʻI 96797
Orders: (800) 468-2800 · Information: (808) 564-8800
Fax: (808) 564-8877
islandheritage.com

ISBN: 0-93154-899-3
First Edition, Second Printing, 2006

Table of Contents

Acknowledgments

Many people have helped with the preparation of this book in one way or another by providing photos, information, and encouragement, and to all I am very grateful. I especially want to thank the following members of Mauna Kea's diverse community: Cheryl Braunstein, Jean-Charles Cuillandre, John Dvorak, Kahalelaukoa Ka'ahanui Ell, Michael Gregg, Don Hall, Catherine Ishida, Laura Kraft, Jon Lomberg, Peter Michaud, Alison Peck, Douglas Pierce-Price, Kimo Pihana, John Rayner, Walter Steiger, Marianne Takamiya, and Christian Veillet. I would also like to thank Allen Hoof of the Hawai'i State Archives, Lisa Martin and Maria Schuchardt of the Lunar and Planetary Laboratory at the University of Arizona, Heather Lindsay of the American Institute of Physics Emilio Segrè Archive, and Gary Pressel of Adtech Photo Imaging for their invaluable assistance in obtaining historical photographs. Thanks also to Lynne Wikoff and Kirsten Whatley for their excellent editorial work. Most important of all, I would like to thank my loving family—my son and best buddy, Caden, for his infinite patience while Daddy was busy writing this book, and my wife, Cheryl, for her patience, support, and sage advice.

Except where indicated otherwise, all photographs are by the author. A few are composite photos intended to capture the feeling of being on Mauna Kea at night.

I would like to emphasize that the views expressed here are my own, and not necessarily those of the astronomical community or the University of Hawai'i.

Finally, it is my hope that all who choose to visit Mauna Kea will do so in a way that is respectful of the Hawaiian culture, the natural beauty, and the scientific endeavors of this unique place.

A Sacred Mountain

Image by Bruce N. Omori.

The Reverend William Ellis, a nineteenth-century missionary to Hawai'i, wrote in 1827 that Mauna Kea "appeared above the mass of clouds that usually skirt the horizon, like a stately pyramid, or the silvered dome of a magnificent temple." Here Mauna Kea (left) and Mauna Loa (right) are seen at sunrise from the neighboring island of Maui. Image courtesy of Walter Steiger.

Mauna Kea kuahiwi ku ha'o i ka mālie.
Mauna Kea, standing alone in the calm.
~ '*Ōlelo No'eau*, Mary Kawena Pukui, collector and translator, Bishop Museum Press

Mauna Kea rises majestically from the depths of the Pacific Ocean as if Earth itself were reaching up to touch the heavens. Measured from its base to its peak, it is the tallest mountain in the world, towering fourteen thousand feet above sea level with another nineteen thousand feet hidden below the ocean's surface. To stand on the summit of Mauna Kea is to stand above almost half of our planet's atmosphere.

Like all the volcanoes that make up the Hawaiian Islands, Mauna Kea was born of fire. Sometime in the distant past, roughly a million years ago, the mountain was created by a volcanic hot spot on the ocean floor that spewed molten lava from Earth's interior. As the newborn Mauna Kea grew, its crest slowly emerged from the ocean's watery womb and continued rising until it pierced the clouds. Carried along by Earth's constantly shifting crust, the massive mountain—which today is bigger than the entire island of O'ahu or Maui—gradually drifted away from its parental hot spot and grew no more. But although Mauna Kea's last volcanic eruption occurred some four thousand years ago, scientists cannot rule out the possibility that it might someday erupt again.

The realm of the gods

For most of its history, Mauna Kea existed in splendid isolation, shaped by numerous episodes of volcanic activity and ice ages to which no human was witness. The first people to behold Mauna Kea were Polynesians, whose legendary ability to sail across the vast Pacific Ocean by observing the stars, winds, waves, and cloud patterns brought them to Hawai'i's shores nearly two thousand years ago. Mauna Kea's immense size and, sometimes, snowcapped peak—unknown sights elsewhere in the tropical Pacific—must surely have amazed them.

ABOVE: Twilight comes to the Big Island, and the lights of Hilo begin to glow beneath the silhouette of Mauna Kea. The mountain plays a major role in the island's weather, as moist trade winds coming from the east rise up Mauna Kea's steep slopes and condense into clouds and rain. Because of this, the eastern side of the mountain is a lush rain forest (Hilo is the rainiest city in the United States, receiving from 150 to 300 inches of rain a year—nearly twice that of Seattle), while the western side of the island experiences desertlike conditions.

LEFT: The Big Island of Hawai'i, photographed in 2002 from the International Space Station as it orbited 245 miles above Earth's surface. The island is dominated by Mauna Loa (left) and Mauna Kea (right). Mauna Kea is just slightly higher than Mauna Loa, and has much steeper slopes. Image courtesy of NASA.

LEFT: Poli'ahu, the goddess of snow, often casts her icy spell across Mauna Kea's landscape during winter. Small cinder cones like this dot the slopes of Mauna Kea. They formed from explosive volcanic eruptions that ejected mounds of cinders and ash into the air.

OPPOSITE PAGE: As the sun sets, Mauna Kea's shadow stretches across Earth's upper atmosphere and the clouds that lie below the mountain's summit.

Perhaps for these reasons, Mauna Kea has long been a sacred place to the Hawaiian people. The mountain is revered as the embodiment of the *piko*, or umbilical cord, that connects Hawaiians back through time to their ancestral origin as descendents of the gods. It is the home of Poli'ahu, the goddess of snow, and her sister Lïlïnoe, the goddess of mist. Some say the white crown of snow that frequently covers its peak during winter months, from November to April, and occasionally during other times of the year gave the mountain its name: Mauna Kea means "White Mountain." Others suggest that Mauna Kea is a shortened form of Ka Mauna a Wākea, or "Wākea's Mountain," since, according to ancient lore, the island of Hawai'i was the firstborn child of Wākea and Papa-hānau-moku, the Sky Father and Earth Mother.

Because of its sacred nature as *wao akua*—the region of the gods and goddesses—only the most elite members of ancient Hawai'i's society were permitted to ascend the summit of Mauna Kea. The many ancient stone shrines and altars sprinkled around Mauna Kea's slopes—nearly one hundred have been found so far—attest to the mountain's powerful cultural and spiritual significance. This holiest site in the Hawaiian Islands was also the final resting place of chiefs, priests, and other high-ranking members of society, whose remains

ABOVE: Pu'u Wēkiu, the highest peak on Mauna Kea. Mark Twain wrote in 1873, "If you want snow and ice forever and ever, and zero and below, build on the summit of Mauna Kea."

were buried in hidden graves. Today, some native Hawaiians continue to worship on Mauna Kea and to honor their cultural heritage there, just as their ancestors did for centuries before them.

While Mauna Kea's summit was a spiritual realm to Hawai'i's early inhabitants, Mauna Kea's lower elevations provided many practical benefits. One of the most important was the Keanakāko'i adze quarry that stretched for several miles around the twelve-thousand-foot level of Mauna Kea's slopes. Because metal was unknown in Hawai'i, for centuries fragments of dense volcanic rock were harvested from the mountainside and shaped into stone tools used to cut down trees from which canoes, weapons, and other items essential for survival were crafted. Although adze quarries were also found on other Hawaiian islands, Mauna Kea was prized for having the highest quality stone from which the finest tools could be made.

A *palila* bird, found only on Mauna Kea. Image by Tom Dove.　A Mauna Kea silversword. Image by Ron Dahlquist.

LEFT: One of the most intriguing residents of Mauna Kea is the *wēkiu* bug. Found solely on the mountain's summit—they take their name from the Hawaiian word for "peak"—*wēkiu* bugs are supremely adapted to life in this harsh environment. With no vegetation and few food sources available, they survive among the rocks and cinders by eating other dead or dying insects blown to the summit by strong upslope winds. Here a *wēkiu* bug makes a meal of an unfortunate fly. Image courtesy of Peter T. Oboyski.

OPPOSITE PAGE: A lone *māmane* tree growing on Mauna Kea's southern slopes. Many of Hawai'i's native plants likely arrived here long ago as seeds eaten and deposited by migratory birds, or carried along by wind or ocean currents from distant lands. The geographical remoteness of the Hawaiian Islands allowed transplanted species of plants and animals to evolve in relative isolation over hundreds of thousands of years. Some developed unique adaptations that have enabled them to survive on Mauna Kea's rugged slopes. The beautiful *palila* bird, for example, is only found among the *māmane* and *naio* trees at Mauna Kea's lower elevations. Unfortunately, destruction of its forest habitat by grazing cattle and sheep has endangered the existence of this species. The same is true of the Mauna Kea silversword, a plant that can live for decades before it flowers once and then dies.

OPPOSITE PAGE: High up the arid slopes of Mauna Kea lies the only mountain lake in the island chain. Lake Waiau, whose name means "swirling water," was carved by glaciers that last covered the mountain's summit area about fifteen thousand years ago. Its waters are replenished by snow and rain runoff from surrounding cinder cones. Ancient Hawaiians believed that the lake contained pure water created by the deities, and ascribed special medicinal and spiritual powers to it. Generations of Hawaiian families have come to Lake Waiau to place their newborn children's umbilical cords in its pristine waters as offerings to the spirit world to assure a long and healthy life.

RIGHT: One of the earliest known photos of Mauna Kea, circa 1892. Image courtesy of Hawai'i State Archives.

BELOW: Lake Waiau, circa 1910. Image courtesy of Hawai'i State Archives.

A simple stone shrine stands on silent guard atop
Mauna Kea. Hawaiian reverence for the mountain is
expressed by a traditional saying: *O Mauna Kea ko
kākou kuahiwi la'a*, "Mauna Kea, our sacred mountain."

LEFT: Despite its location in the tropics, Mauna Kea is so tall that it experienced ice ages at least four times in its history, the last about fifteen thousand years ago. It is the only mountain in Hawai'i where glaciers once existed.

BOTTOM LEFT: An enormous pile of discarded stone chips in the Keanakāko'i adze quarry on the upper slopes of Mauna Kea. Archaeological evidence suggests that this quarry was mined as early as AD 1100 for its prized basaltic stone and was in continuous use for nearly seven hundred years. Stone "blanks" were chipped from rock outcroppings and carefully shaped into tools used for carving wooden canoes, weapons, and other implements used in daily life. Although adze quarries also existed elsewhere in the Hawaiian Islands, Mauna Kea was renowned for producing the strongest adzes, thanks to the unusually dense volcanic rock that formed there when lava flows occurred under glaciers that covered the mountain's summit region during past ice ages. Stone from the Keanakāko'i quarry was considered so valuable that early Hawaiians were willing to labor for long periods under extreme conditions on the barren landscape near Mauna Kea's peak, sleeping in lava caves and enduring freezing temperatures at night.

BOTTOM RIGHT: Some adze fragments found on Mauna Kea. Master craftsmen used the small round stones at the top to shape basalt harvested from the mountainside into tools. The larger stone fragments at the bottom are unfinished adzes.

The realm of the stars

"The ancient Hawaiians were astronomers," wrote Hawaiʻi's last reigning monarch, Queen Liliʻuokalani, in 1897. Indeed, when Captain James Cook and his ships arrived on Hawaiʻi's shores in 1778, he was met by a people with a long tradition of watching the heavens. The starry skies guided the Hawaiians not only as they sailed across the vast Pacific Ocean but also in their daily lives. Crops were planted, fish were caught, wars were fought, and religious festivals were celebrated according to the phases of the moon and the seasonal positions of the stars in the sky. Hawaiian astronomers, called *kilo hōkū*, or "star watchers," were among the most esteemed members of society.

LEFT: One of Hawaiʻi's most beloved kings was David Kalākaua, the "Merrie Monarch," who reigned from 1874 to 1891. Keenly interested in astronomy, King Kalākaua wished to establish an observatory in Hawaiʻi. He succeeded in bringing the first permanent telescope to Hawaiʻi in 1883. It was a small, 0.1-meter (5-inch) telescope imported from England that was housed at the private Punahou School in Honolulu. Image courtesy of Hawaiʻi State Archives.

OPPOSITE PAGE: The ancient and the modern coexist on Mauna Kea. A traditional Hawaiian altar on Puʻu Wēkiu, the highest point on Mauna Kea, is seen here, with two astronomical observatories visible in the background.

Sadly, much of the ancient Hawaiian knowledge of the heavens has been lost, a casualty of the cultural upheaval that followed European contact. With no written language, Hawaiians passed down their astronomical knowledge orally from generation to generation. Although the Hawaiians had names for hundreds of stars and constellations, the nineteenth-century mission-

Astronaut Lacy Veach, who made Hawai'i his home, took the photo here aboard the Space Shuttle in November 1992. It shows an ancient stone adze from Mauna Kea's Keanakāko'i quarry floating freely in the shuttle flight deck while the island of Hawai'i is seen 160 miles below through the window. The adze was given to Veach by his grandfather. Image courtesy of NASA.

aries who first transcribed the Hawaiian language had little or no knowledge of astronomy themselves, and so they often recorded Hawaiian star names without identifying which stars they represented. More than three hundred Hawaiian star names have survived to the present, but less than half are associated with specific stars.

No one knows for certain whether the ancient Hawaiians made astronomical observations from Mauna Kea. Perhaps the *kilo hōkū* came to the tallest mountain in the Pacific to discern signs in the starry skies from its vantage point closest to the heavens. Or it may be that the summit was considered too divine for such human activity.

The first telescope in Hawai'i arrived and departed with Captain Cook in 1778: his ships' inventory listed four telescopes. A century would pass before Hawai'i's King Kalākaua brought

The top of Mauna Kea is home to a multinational village of telescopes, the tools of modern astronomical explorers. Mauna Kea's summit is considered by many astronomers to be the best place in the world to explore the cosmos. Image by Joe D'Amore.

OPPOSITE PAGE: As the day ends and the soft light of sunset gives way to the encroaching darkness, the sky is transformed into a breathtaking display of countless sparkling stars. The Milky Way arches overhead as if the goddess of snow, Poli'ahu, had flung snowflakes across the heavens. Here the Gemini North Telescope is seen as it looks out into the distant universe. Image courtesy of Peter Michaud and Kirk Pu'uohau-Pummill/Gemini Observatory.

BELOW: As the sun sets over Mauna Kea, the Subaru Telescope prepares for the night ahead.

the first permanent telescope to the islands in 1883, and almost another century would pass before the first telescope was built on Mauna Kea.

Today, thirteen of the biggest and most sophisticated telescopes ever created stand near the summit of White Mountain. In a sense they are modern testaments to the glory of Mauna Kea, built by astronomers who, like the Hawaiians who came here before them, revere the mountain as a gateway to the heavens. The same curiosity to learn what lies beyond the horizon that inspired the ancient Polynesians to set sail for new lands and brought them to Hawai'i so long ago inspires astronomers today to search the cosmic ocean to learn about distant stars, distant galaxies, distant worlds.

From one of the youngest spots on Earth, a new generation of explorers looks up into the black night speckled with ancient starlight and wonders how this all came to be.

The Road to Mauna Kea

Image by Bryan Lowry.

The road to Mauna Kea intersects with the Big Island's Saddle Road, which winds through mile after mile of lava fields.

Kūlia i ka nuʻu.
Strive for the summit (motto of Hawaiʻi's Queen Kapiʻolani, who lived from 1834 to 1899).
~ʻŌlelo Noʻeau, Mary Kawena Pukui, collector and translator, Bishop Museum Press

Gerard Kuiper had a dream. The Dutch-born astronomer, who was head of the Lunar and Planetary Laboratory at the University of Arizona, wanted to find the best astronomical site in the world, a place where the nights are clear and the starry skies are breathtaking. There he would build an observatory to capture the most magnificent views of the heavens obtainable anywhere on Earth. Kuiper's quest brought him to Hawaiʻi in 1963.

Haleakalā, the highest peak on the island of Maui, was Kuiper's first choice. A few years earlier the University of Hawaiʻi had built a small observatory there to study the sun because daytime viewing conditions were known to be excellent. Nighttime tests by Kuiper and his assistants confirmed that astronomical images of extraordinary clarity could be obtained from

Haleakalā. But Kuiper found, to his disappointment, that clouds and fog often swept over Haleakalā at night, blocking the skies. So he turned his attention to the neighboring island of Hawai'i, where the peaks of Mauna Kea and Mauna Loa—both almost a mile higher than Haleakalā's—could often be seen poking through the clouds into the clear skies above. Because Mauna Loa was still an active volcano, and hence a risky place to build an observatory, Kuiper chose Mauna Kea.

There was just one problem: the summit of Mauna Kea was inaccessible except by foot, horse, or mule along narrow trails that meandered along the mountain's gnarled slopes. To build an observatory on Mauna Kea, a road would be needed to transport equipment up the mountain. So in January 1964 Kuiper flew to Honolulu to meet with Governor John A. Burns to ask the State of Hawai'i to finance

Dutch-born astronomer Gerard Kuiper was one of the leading scientists of the twentieth century, a pioneer in the study of our solar system, and a driving force behind the development of astronomy on Mauna Kea. Image from AIP Emilio Segrè Visual Archives, *Physics Today* Collection.

Mauna Kea's steep slopes provide an amazing diversity of environments. Mark Twain, who visited Hawai'i as a young reporter in 1866, once remarked that from the summit of Mauna Kea, "you can look down upon all the climates of the earth." This photo was taken near the summit, looking down toward the lush green valley that lies between Mauna Kea and Mauna Loa.

construction of a six-mile-long jeep trail. Burns quickly agreed; the Big Island was still reeling from a tsunami that had devastated much of downtown Hilo four years earlier, and Burns hoped that astronomy might provide a new and sustainable industry to help rebuild the local economy.

Construction of a road to the summit of Mauna Kea began in April 1964 and was completed in a few weeks. Two months later, with funding from NASA, a small prefabricated observatory with a 0.3-meter (12-inch) telescope inside was placed on Pu'u Poli'ahu, one of the highest peaks on the mountain. Nighttime observations exceeded expectations; Kuiper's efforts had indeed paid off. In July of that year he declared enthusiastically, "This mountaintop is probably the best site in the world—I repeat in the world—from which to study the moon, the planets and the stars. . . . It is a jewel! This is the place where the most advanced and powerful observations from this Earth can be made."

OPPOSITE PAGE: A photo of the snowcapped summit of Mauna Kea taken in 2003 from the International Space Station as it orbited 247 miles (398 km) above Hawai'i. Image courtesy of NASA.

ABOVE: Snowcapped Mauna Kea seen from the tropical splendor of Hilo Bay.

And so began the era of modern astronomical exploration on Mauna Kea. NASA, which planned to provide funds to build and manage a major new telescope on the mountain, received bids from Kuiper's home institution of the University of Arizona, Harvard University, and the University of Hawai'i. Although both Arizona and Harvard had much more established and pre-eminent astronomy programs at the time, NASA chose the fledgling astronomy group at the University of Hawai'i. Kuiper was bitterly disappointed by the decision, as it meant that he would not realize his dream of building a telescope on Mauna Kea, and felt that his discovery of the mountain as a superlative astronomical site had been stolen from him.

An aerial view of the summit of Mauna Kea, circa 1930. Although history has not recorded the first Hawaiian to ascend the mountain, the first non-Hawaiian to reach its summit was the Reverend Joseph Goodrich, who hiked alone under the light of a full moon on August 25, 1823. Image courtesy of Hawai'i State Archives.

The University of Hawai'i's Institute for Astronomy was created in 1967 to manage the observatories on both Mauna Kea and Haleakalā. Construction of the University of Hawai'i's 2.2-meter (88-inch) Telescope also began in 1967 with funding from NASA, and when com-

30

pleted in 1970 it was the seventh largest in the world. With additional funding from NASA, the University of Hawai'i also erected a telescope with a 0.6-meter (24-inch) diameter mirror on the mountain in 1968. Mauna Kea's reputation as a superb astronomical site quickly attracted international interest from as far away as Europe and Asia, and other telescopes soon followed—three in 1979, two more in 1986, and another six during the 1990s.

Today, there are more major telescopes—thirteen—on Mauna Kea than on any other mountain peak in the world, with nearly one billion dollars invested in this international village by Argentina, Australia, Brazil, Canada, Chile, France, Japan, the Netherlands, Taiwan, the United Kingdom, and the United States. From Kuiper's dream, Hawai'i has grown to become one of the leading centers of astronomical research and education in the world today. University of Hawai'i astronomers are in the enviable position of receiving a guaranteed 10 to 15 percent share of time on all telescopes on Mauna Kea in return for allowing them to be built in Hawai'i. Such access to the world's most powerful collection of telescopes has helped the university's Institute for Astronomy become a leading center of astronomical research. Astronomy has also become a major industry in Hawai'i, an economic engine that pumps millions of dollars into the local economy and employs hundreds of people.

Astronomer Alika Herring, who grew up in Hawai'i, sits in the first observatory on Mauna Kea in 1964. Herring, a renowned telescope maker who built the small telescope seen here, did most of the site testing on Mauna Kea as part of Gerard Kuiper's team. He considered it the best astronomical site he had ever seen. Image courtesy of the Lunar and Planetary Laboratory, University of Arizona.

RIGHT: From the current astronomers' residence halfway up the mountain, the road to the summit of Mauna Kea is ten miles long, much of it steep and winding. The first five miles are unpaved and very rough in places; road crews must resurface it weekly. The road is paved near the summit to minimize dust blowing into the telescopes.

BELOW: Their enormous sizes sometimes make it easy to forget that the telescopes on Mauna Kea are delicate pieces of precision equipment. Here the mirror for the Canada-France-Hawai'i Telescope (CFHT) is being transported to the summit in 1978. Because even the slightest bump might damage the delicate mirror, it is transported at a snail's pace up the winding gravel road to the summit, as tractors and workers in front clear the path. Image courtesy of Canada-France-Hawai'i Telescope.

Never one to rest on his laurels, Gerard Kuiper always held out hope that an even better astronomical site than Mauna Kea might exist in some remote corner of the world, and he continued what became a lifelong quest to find it. But he never did. Kuiper died in Mexico in 1973 at the age of sixty-eight while scouting possible sites for new observatories there. The telescopes on the summit of Mauna Kea remain his legacy.

"A sky wonderful with stars"

Gerard Kuiper was not the first person to be awestruck by the splendor of the night sky over Mauna Kea. One of the first non-Hawaiians to explore Mauna Kea—a man who eventually died there—was the famous Scottish botanist David Douglas, for whom the Douglas fir tree is named. Arriving at the summit of the mountain in January 1831, Douglas wrote in his journal, "Never, even under a tropical sky, did I behold so many stars," which he described as shining "with an intense brilliancy."

Others who followed in Douglas's footsteps were equally impressed. Recounting her ascent of Mauna Kea in 1873, Victorian-era travel writer Isabella Bird observed, "The mist as usual disappeared at night, leaving a sky wonderful with stars."

What makes Mauna Kea such a splendid place from which to view the heavens? In three words: location, location, location.

We live at the bottom of an ocean of air. After journeying across billions or trillions of miles of mostly empty space, light from distant stars and galaxies must plow its way through Earth's atmosphere to reach telescopes on the ground. Unfortunately, airplanes are not the only things that get bounced around by air turbulence; so does starlight. Swirling, turbulent currents in Earth's atmosphere jostle starlight as it passes through; this is what causes the stars to appear to twinkle. For astronomers, such twinkling means blurry views of planets, stars, and galaxies, which make details harder to see. Without our planet's atmosphere, the stars would look like brilliant pinpoints of light rather than shimmering orbs.

ABOVE: Mauna Kea's harsh conditions make it a challenging place to work. Thousands of gallons of water must be trucked up the parched mountain each week to supply the facilities there; rock falls must be removed; and snowfalls, which can often be several feet deep, must be cleared.

LEFT: As the last traces of vegetation vanish above an elevation of about eleven thousand feet, Mauna Kea's barren landscape takes on an otherworldly appearance. So inhospitable is the environment on the mountain's upper slopes that scientists have used it to study conditions similar to those on other planets. A 1997 scientific expedition, for example, searched for bacteria living in the soil on Mauna Kea's summit because conditions there may be similar to those on Mars billions of years ago when the now lifeless planet is believed to have been warmer and wetter. By studying the range of conditions in which life can survive on Earth, scientists hope to learn about the possibility of life existing on other planets.

ABOVE: The mile marker six sign on the summit road after a heavy snowfall.

OPPOSITE PAGE: Howling winds, freezing temperatures, and occasional blizzards are not the Hawai'i most people think of, but these conditions are encountered on Mauna Kea's summit—the only place in Hawai'i that has a snowplow. Sudden storms can quickly engulf the summit in blizzard conditions during winter months. When that happens, the summit must be evacuated immediately to ensure that nobody is trapped. This shows a whiteout on Mauna Kea, with the James Clerk Maxwell Telescope (JCMT) barely visible through the mist.

LEFT: Sometimes just getting to work can be a challenge on Mauna Kea.

Dangers abound on Mauna Kea. Here a sign warns travelers to beware of cows crossing the road at night at lower elevations.

BEWARE OF
INVISIBLE COWS

MOST OF THE MAUNA KEA ACCESS ROAD BELOW HALE POHAKU IS OPEN CATTLE RANGE, AND THE COWS FREQUENTLY CROSS THE ROAD. DARK COLORED COWS ARE OFTEN INVISIBLE IN DARKNESS AND/OR FOG. USE EXTREME CAUTION AND DRIVE VERY SLOWLY IN THIS OPEN RANGE.

"The solitude and silence—how deathlike everything is! No sound is heard, no living creature stirs." Letter from Charles de Varigny describing his impressions on reaching the summit of Mauna Kea in 1857.

At its height of nearly fourteen thousand feet above sea level, the summit of Mauna Kea rises above 40 percent of our planet's atmosphere, allowing telescopes there to reach up to collect light from stars and galaxies before the thicker layers of air below distort it. The other 60 percent of the atmosphere that remains overhead is exceptionally stable, thanks in large part to cleansing winds that blow unimpeded across thousands of miles of open ocean to reach Hawai'i. The result is a site that consistently yields the sharpest astronomical images obtainable anywhere on Earth, remarkably free of the atmospheric distortions that usually blur astronomers' views of the heavens from most other locations on our planet.

Good weather is another important factor; there is no point building telescopes in places where skies are frequently cloudy. A tropical inversion layer that is usually located over the island of Hawai'i traps clouds well below the summit, keeping the sky overhead clear and dry most of the time. Mauna Kea has one of the highest percentages of clear nights of any astronomical site in the world. Typically, about 80 percent of nights each year are usable for astronomical observations.

Finally, there is the human element. As anyone who lives in a big city knows, the cumulative glare from streetlights, cars, and buildings illuminates the night sky and can make it

Its geographical isolation—over two thousand miles of open ocean to the nearest continent—is one reason Mauna Kea is such a superb site for astronomical observations. With no nearby landmasses to impede air currents, winds flow smoothly across the mountain's peak, resulting in less air turbulence overhead and greatly reduced atmospheric distortions. Consequently, Mauna Kea is renowned for consistently producing the sharpest astronomical images obtained by telescopes anywhere on Earth. This 1997 photo shows the Big Island of Hawai'i and the rest of the Hawaiian island chain as seen from the Space Shuttle. Image courtesy of NASA.

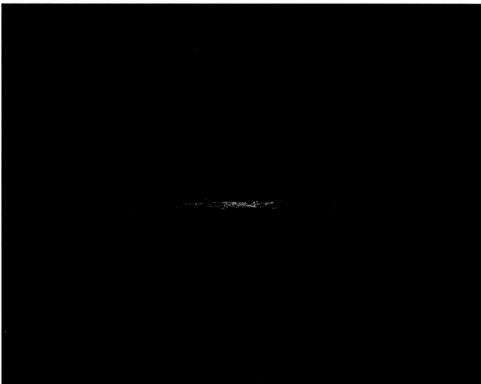

The largest source of light pollution on Mauna Kea is the town of Hilo (population forty-five thousand), seen here at night from the mountain's top. Fortunately, at present Hilo's lights have only a minimal impact on astronomical observations.

ABOVE: Mauna Kea's peak is usually high above the clouds. Occasionally, however, the clouds rise up to engulf the summit. Here the domes of the Canada-France-Hawai'i, Gemini North, and University of Hawai'i telescopes can be seen rising through the mist as sunlight warms the previous day's snowfall. Image by Joe D'Amore.

OPPOSITE PAGE: The summit of Mauna Kea offers spectacular panoramic views of the island of Hawai'i, the Pacific Ocean, and the cosmos.

difficult to see all but the brightest stars. This light pollution is a problem that plagues professional and amateur astronomers around the globe. Fortunately, the Big Island of Hawai'i's low population density—imagine an area the size of Connecticut with only 4 percent of that state's population—means that Mauna Kea is surrounded by very little artificial light. The absence of a nearby metropolis together with local regulations that restrict light pollution on the island have resulted in some of the darkest skies to be found anywhere in the world.

On a moonless night the skies over Mauna Kea are so dark that the glow of the Milky Way itself casts shadows across the landscape. It is truly a sublime sight.

A Gentle Rain of Starlight

RIGHT: The setting sun casts long shadows across Mauna Kea. Here the silhouettes of the Canada-France-Hawai'i, Gemini North, and University of Hawai'i telescopes can be seen against the backdrop of one of Mauna Kea's many cinder cones.

OPPOSITE PAGE: As night comes to Mauna Kea, telescopes open their eyes to the sky. The stars appear as trails of light in this ten-minute exposure, a feature that is caused by Earth's rotation. The color of each star's light reveals its temperature; blue stars are hottest and red stars are coolest.

O na hoku no na kiu o ka lani.
The stars are the spies of heaven.
~ *'Ōlelo No'eau*, Mary Kawena Pukui, collector and translator, Bishop Museum Press

Light from distant planets, stars, and galaxies falls softly onto Earth's surface every second of the day and night. Collecting this light to learn about these faraway worlds is what astronomy is all about.

Through most of history, astronomy was done with the naked eye. But the vast majority of light that comes to us from the distant universe is too dim for the human eye to see. The invention of the telescope in the 1600s allowed astronomers to see farther and more clearly into space than ever before. What they saw changed forever our view of the cosmos and our place in it. In a single year—1610—the famous Italian scientist Galileo Galilei

Light

What do a rainbow's dazzling colors, radio and television broadcast signals, microwaves used to cook food, and X-rays that reveal bones beneath flesh all have in common? They are all different forms of the same thing—light. Astronomers recognize a number of different kinds of light, ranging from most to least energetic:

gamma rays
X-rays
ultraviolet
visible
infrared
submillimeter
radio

Most of these are invisible to the human eye, but not to telescopes.

Light is a form of energy, and different types of light possess different amounts of energy. Gamma rays, for example, are the most energetic kind of light—so energetic that they can damage or destroy cells. Radio waves, on the other hand, continuously bounce off our bodies without any harmful effects because they have very little energy. The sun is an abundant source of visible light, so it is probably no coincidence that our eyes have evolved to see visible light energies, since the reflection of sunlight off objects around us allows us to perceive them.

OPPOSITE PAGE: Mauna Kea's summit is home to more major telescopes than any other place on Earth. Why are so many telescopes needed? More telescopes mean more knowledge about the vast universe that remains unexplored. There are more than one hundred billion stars in our Milky Way galaxy alone, and more than one hundred billion other galaxies scattered throughout the universe. To collect enough light to learn about each of these objects requires an enormous amount of telescope time; even just counting to one hundred billion at a rate of one number per second would take a person more than three thousand years. The diversity of telescopes and instruments on Mauna Kea also allows astronomers to collect and analyze the various kinds of light emitted by astronomical objects, each of which provides different information. Here the Gemini North, University of Hawai'i 2.2-meter (88-inch) and United Kingdom Infrared telescopes are illuminated by the glow of the full moon as they collect light from distant stars and galaxies.

Like all telescopes, NASA's Infrared Telescope Facility, shown here, is a time machine that allows astronomers to look not only far away, but long ago, too. Because light takes time to travel through space, we see a distant star or galaxy as it appeared in the past, when its light first left on its journey through space. Each photon of light that enters the telescope ends a journey that may have begun thousands, millions, or even billions of years ago. In some cases the objects might not even exist anymore, or might now be in different locations.

discovered that the moon has mountains and craters, Jupiter has moons, Saturn has rings, and the faint, misty swath of light known as the Milky Way is made of countless stars like our sun. And he did all this using a simple, homemade telescope. But Galileo could never have imagined the telescopic marvels that would someday grace the summit of Mauna Kea, or the secrets of the heavens they would reveal.

Telescopes are housed in protective domes to shield them from wind, rain, sunlight, and dust. A small slit opens to allow the telescope to view the sky. Because temperature differences of even just a few degrees between inside and outside the dome can create pockets of air turbulence that blur images of stars and galaxies, domes are ventilated to allow air to circulate freely through them. A coat of white paint or other reflective material prevents sunlight from being absorbed and heating up the interior of the dome during the day. To minimize temperature differences, the interiors of many domes are also cooled during the day to the expected nighttime temperature.

A face-on view of the Keck I Telescope's mirror, which spans 10 meters (33 feet) from end to end. Telescope mirrors must be shaped and polished to near perfection to focus light properly. A very thin layer of metal—no more than a fraction of the thickness of a human hair—coats the glass to produce a reflective surface. Making enormous telescope mirrors with precise shapes is a time-consuming process, usually requiring several years. The Keck I and II telescopes use a novel mirror design, tiling together thirty-six smaller, easier-to-manufacture mirrors, each just 1.8 meters (6 feet) across. These telescopes have seventeen times the light-gathering area of the Hubble Space Telescope but cost only a fraction of what Hubble did. Image courtesy of W. Scott Kardel and W. M. Keck Observatory (2004).

Mauna Kea's giant eyes on the sky

Astronomers use telescopes to gather as much light as possible from a region of the sky and bring it into focus. This light is extremely faint—a lightbulb gives off more light in a single second than all the starlight the telescopes on Mauna Kea have collected over the past forty years.

At the heart of each Mauna Kea telescope is a curved mirror that acts like a bucket to collect droplets of starlight—photons—that rain down from the sky. Depending on the type of light being collected, this mirror may be made of glass coated with a microscopically thin layer of reflective aluminum, or it may be made entirely of metal.

Just as a big bucket collects more raindrops than a small one, a big telescope mirror collects more photons than a small one. The bigger the mirror, the brighter and sharper the images it produces. To gather enough light to see the faint flickers of stars and galaxies at the farthest reaches of the universe requires very large mirrors. The twin Keck telescopes, for example, use giant mirrors 10 meters (33 feet) across to see objects millions of times fainter than the unaided human eye can see. Their light-collecting power is so great that they could detect the faint glow of a lightbulb on the moon.

The thirteen telescopes on Mauna Kea today provide a snapshot of the progress in telescope design over the past four decades. Many of the first telescopes on the mountain, such as the University of Hawai'i and Canada-France-Hawai'i telescopes, had thick, rigid mirrors to prevent the heavy glass from sagging under its own weight and distorting the reflected light. This limited their size but not their ability to make countless important

The heart of a telescope is its light-collecting mirror. Tons of massive machinery and complex electronics are used to deposit a microscopically thin layer of aluminum on the mirror. Here a worker inspects the Gemini North Telescope's mirror after it received its first coat of reflective aluminum in 1998. A mirror must be recoated from time to time as its reflective surface tarnishes and collects dust.

To focus light accurately the mirror's shape must also be as close to perfect as possible. Gemini's mirror of 8 meters (26 feet) in diameter, which collects more light than a million pairs of human eyes, is so smooth that if it were expanded to the size of Earth, the largest bump would be only a few inches high. Image courtesy of Gemini Observatory.

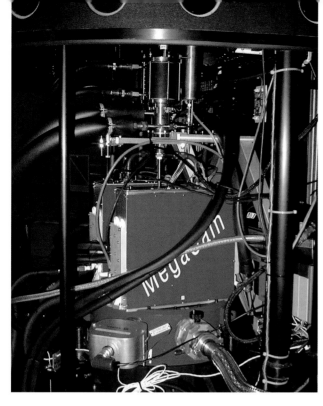

astronomical discoveries, then and today.

For years, giant mirrors were not feasible because of the difficulty and expense of manufacturing massive slabs of glass weighing many tons and mounting them securely in telescopes. However, in the 1990s, two revolutionary mirror designs changed all that. One innovation, used by both the Gemini and Subaru telescopes, was to use a wider but thinner, more lightweight mirror supported by hundreds of tiny computer-controlled pistons that prop up the glass and continuously nudge it back into perfect shape. An altogether different approach, used by the two Keck telescopes, was to replace the traditional mirror made from a monolithic piece of glass with a mosaic of numerous smaller mirror segments that work together to collect light, as if a single large mirror had been cut into smaller pieces with a cookie cutter. Thanks to these advances in mirror size, the latest generation of telescopes on Mauna Kea has roughly five times the light-collecting power of the previous one—and many thousands of times more than Galileo's original telescope.

More than meets the eye

Telescopes not only collect far more light than the human eye, they also collect different kinds of light that the human eye is incapable of seeing. Optical telescopes—those that harvest visible light—are just one of the tools used by astronomers to explore the universe.

The popular image of an astronomer squinting through a telescope eyepiece is a relic of the past. Today's professional astronomers almost never look through telescopes because the human eye detects only a tiny fraction of the light it receives. This problem is even more severe on Mauna Kea's peak, where the lack of oxygen dramatically reduces night vision. Most astronomical observations are made with electronic detectors that are much more sensitive than the human eye. The images are then stored, displayed, and analyzed on powerful computers. The Canada-France-Hawai'i Telescope boasts the world's largest astronomical camera, called MegaCam (shown here). It has 340 million pixels—about one hundred times the number found in a typical consumer digital camera—which yield remarkable views of the heavens. Building state-of-the-art instruments such as MegaCam is a slow and expensive process, often requiring several years of effort. Image courtesy of Christian Veillet and the Canada-France-Hawai'i Telescope Corporation.

"Come quickly, I am tasting the stars!"

Legend has it that after his first sip of champagne, the seventeenth-century French monk Dom Pérignon exclaimed, "Come quickly, I am tasting the stars!" Unfortunately, astronomers cannot taste or touch stars and other celestial objects they wish to study. For this reason, they have become experts at extracting information from the light that comes from these distant objects.

The image here shows the light from a star as seen with the Keck II Telescope. A special instrument on the telescope, called the Echellette Spectrograph and Imager, acts like a high-tech prism to disperse the star's light into its component colors. Patterns of lines visible in this faint glimmer of starlight reveal a wealth of information about the object, such as what the star is made of, how hot it is, and how fast it is moving through space.

Stars, galaxies, planets, quasars, and other celestial objects send out a symphony of light, including the visible light that our eyes can see and many others kinds that they cannot, such as X-ray, radio, ultraviolet, infrared, and submillimeter light. If astronomers studied only the visible light that comes from space, they would miss much of the picture; some of the most interesting objects in the universe shine most brightly in light that our eyes cannot perceive. Newborn stars, for example, are often shrouded by clouds of gas and dust that block their visible light from reaching Earth but allow the infrared light they emit to pass right through. Each different kind of light provides a different view of the cosmos.

To make the invisible universe visible, astronomers use specially designed telescopes and instruments. Just as a photograph records ordinary visible light, other types of detectors record other kinds of light. For example, the Very Long Baseline Array (VLBA) telescope located on Mauna Kea uses a large metal dish that acts like a mirror to gather radio waves from space and convert them into a "radio image" of an object. Similarly, the United Kingdom Infrared Telescope (UKIRT) is designed to collect the infrared light that lies just beyond what our eyes can see.

Unfortunately for astronomers, Earth's air hinders observations of this invisible universe. An infrared or submillimeter photon that has traveled for millions or billions of years from a distant star or galaxy may be snuffed out just before it reaches Earth, swallowed up by water vapor or carbon dioxide in our planet's atmosphere before it can reach telescopes on the ground that eagerly await its arrival. Mauna Kea's high altitude and dry air—its peak lies above 97 percent of the obscuring water vapor in our planet's atmosphere—make it ideal for infrared and submillimeter observations, allowing telescopes perched there to snare these photons from stars

Although the telescopes on Mauna Kea do not all look alike, they all use a mirror of some sort to reflect incoming light rays to a detector, which then records the data. Here an intricate mesh of metal rods helps the dish of the James Clerk Maxwell Telescope maintain its shape.

and galaxies before the thicker moist layers of air below have a chance to block them. For this reason Mauna Kea is home to some of the most powerful infrared and submillimeter telescopes in the world. More than half the telescopes on Mauna Kea spend at least part of their time collecting infrared light. Moreover, the four largest telescopes on the mountaintop—the James Clerk Maxwell Telescope, the Caltech Submillimeter Observatory, the Very Long Baseline Array, and the Submillimeter Array—are all designed to collect light that our eyes cannot see directly.

The intense glare of the sun prevents the visible light of planets, stars, and galaxies from being seen during the daytime. However, other types of light, such as radio waves, can still be observed because they are not drowned out by the sun's light. This allows radio telescopes to operate around the clock. Here the Very Long Baseline Array (VLBA) observes radio emissions from a distant galaxy during daylight hours.

As astronomy and technology continue to advance, new and more powerful telescopes will eventually replace older ones. Plans are under way to link Mauna Kea's three submillimeter telescopes together to create a supersized telescope that will produce incredibly detailed images. Also on the drawing board are optical and infrared telescopes with mirrors 30 meters (100 feet) or more in size, an enormous leap forward in light-gathering power. Such giant telescopes will likely have billion-dollar price tags, more than the combined cost of all the telescopes currently on Mauna Kea, but the views they provide of the heavens will be priceless.

Just as astronomical research continues to expand on Mauna Kea, there is growing awareness of the need to protect the mountain's fragile ecosystem, and for increased sensitivity to native Hawaiian cultural concerns over the continued development of Mauna Kea's sacred summit. Achieving a balance between these sometimes conflicting interests may not be easy, but it is essential to ensure that all feel welcome on Mauna Kea to worship as they wish, to practice their cultural heritage, and to study the stars.

When we open our eyes underwater, we discover that things in that environment look blurry and out of focus. Astronomers have a similar experience as they look out into space through the ocean of air above us. Swirling, turbulent motions in our planet's atmosphere jostle light rays, blurring images of distant stars and galaxies. One reason that astronomers come to Mauna Kea is to get above much of Earth's atmosphere and take advantage of the relatively turbulence-free air over the mountain's summit. Thanks to technology, it is now possible to improve even further on Mauna Kea's natural assets. One of the most exciting developments in astronomy in recent years has been the use of "adaptive optics," which allows telescopes to eliminate atmospheric effects by reflecting the collected light off an additional small, bendable mirror whose shape is changed many times per second to counteract distortions caused by Earth's atmosphere. Thanks to adaptive optics, several telescopes on Mauna Kea can now produce razor-sharp images that rival those taken with the Hubble Space Telescope orbiting high above the blurring effects of Earth's atmosphere. Compare the images of a star field taken by the Gemini North Telescope without (top) and with (bottom) adaptive optics. Image courtesy of Gemini Observatory.

Gemini North Telescope

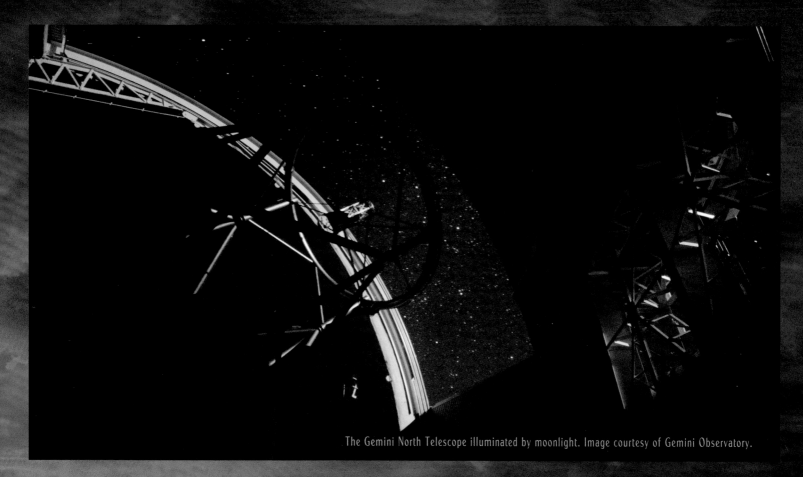

The Gemini North Telescope illuminated by moonlight. Image courtesy of Gemini Observatory.

Named after the fabled twins of Greek mythology, the Gemini Observatory consists of two identical telescopes. One is located near the summit of Mauna Kea; its sibling is in the Andes Mountains of Chile. Together they provide seamless coverage of the northern and southern skies. Both telescopes have mirrors 8 meters (26 feet) across and are designed to collect visible and infrared light. The Frederick C. Gillett Gemini Telescope, usually called Gemini North, saw its first starlight from Mauna Kea in 1999. Gemini South followed in 2001.

Gemini Observatory is a leader in the use of adaptive optics technology, which allows the twin telescopes to see astronomical objects in incredibly fine detail. A human being with the same keen vision as the Gemini North Telescope would be able to see the eyelashes on a person standing ten miles away.

The enormous cost and technological challenges of building cutting-edge telescopes such as Gemini have fostered scientific partnerships between nations. Gemini Observatory is a joint venture between the astronomical communities of the United States, the United Kingdom, Canada, Brazil, Argentina, Australia, and Chile. Astronomers from the partner countries are allotted observing time in proportion to their nations' financial contribution to building and operating the telescopes.

NASA's Infrared Telescope Facility

NASA's Infrared Telescope Facility (IRTF) is one of two telescopes on Mauna Kea dedicated primarily to infrared observing; the other is the United Kingdom Infrared Telescope (UKIRT). Completed in 1979, the IRTF's main purpose is to provide ground-based observations in support of NASA missions to explore our solar system. For example, the IRTF made key observations of the size and rotation of Comet Wild 2 before NASA's Stardust spacecraft flew within 149 miles of the comet's nucleus in 2004 to scoop up samples of comet dust to return to Earth for study.

The telescope, which is operated by the University of Hawai'i's Institute for Astronomy on behalf of NASA, has a mirror 3 meters (10 feet) in diameter. Because it collects infrared light, which does not get as lost in the sun's glare as visible light does, the IRTF can observe some objects such as comets and planets even during the daytime.

One of the IRTF's advantages over many other telescopes is that its instruments can be changed quickly. At most observatories, instrument changes can take many hours, so instruments are usually scheduled to be on telescopes for weeks at a time. At the IRTF, however, they can be changed in as little as fifteen minutes. This makes very efficient use of the telescope by allowing multiple projects to be accommodated in the same night.

James Clerk Maxwell Telescope

Named in honor of the great nineteenth-century Scottish scientist who first explained the physical nature of light, the James Clerk Maxwell Telescope (JCMT) is the largest single telescope in the world designed to collect submillimeter light from space. Built in 1987 as a partnership between Canada, the United Kingdom, and the Netherlands, the JCMT has helped to revolutionize astronomers' understanding of the cold and dusty material that lies between the stars. Astronomers from the partner countries receive about 80 percent of the observing time with the telescope. Another 10 percent goes to University of Hawai'i astronomers, and the remaining time is available to other astronomers around the world.

The JCMT collects submillimeter light using a reflective dish that is 15 meters (50 feet) across and weighs seventy tons. It is made from 276 aluminum panels that can be tilted individually to maintain the mirror's precise shape. To protect it from the wind, rain, dust, and direct sunlight that could damage its sensitive instruments, the opening of the telescope's dome is shielded by the world's largest piece of Gore-Tex, a waterproof fabric also used to make outdoor gear. This material is transparent to submillimeter light, so it does not interfere with observations.

United Kingdom Infrared Telescope

The United Kingdom Infrared Telescope at sunset.

Although many of the telescopes on Mauna Kea spend a portion of their time collecting infrared light, only two are dedicated solely to infrared observations— the United Kingdom Infrared Telescope (UKIRT) and NASA's Infrared Telescope Facility (IRTF).

With a mirror 3.8 meters (12 feet) in diameter, UKIRT is, in fact, the largest telescope in the world devoted exclusively to infrared observations. It is funded by the British government and operated by the Joint Astronomy Centre based in Hilo.

Built as a low-cost observatory, UKIRT was ahead of its time when it was completed in 1979. Unlike the thick, heavy mirrors of most telescopes of that era, UKIRT's mirror is relatively thin and lightweight, weighing less than half as much as other mirrors of comparable size. Its dome is also considerably more compact than those of its contemporaries. Continual upgrades to the telescope and its instruments over the past quarter century have steadily enhanced UKIRT's performance. Today it is regarded as one of the world's finest infrared telescopes, providing unparalleled views of the cosmos. It is also one of the most scientifically productive telescopes in the world; over eighty research papers are published annually based on observations done with UKIRT.

Caltech Submillimeter Observatory

The first submillimeter telescope on Mauna Kea, the Caltech Submillimeter Observatory (CSO) began observing the cosmos in 1986. Its futuristic-looking metallic dome houses a metal dish that spans 10.4 meters (34 feet). Aluminum panels on the dish's curved surface reflect submillimeter light and focus it onto sensitive detectors. CSO is used to study some of the coldest regions of the universe, providing information on such things as the abundance of different types of molecules in outer space, the birth and death of stars, and the evolution of galaxies. CSO is the only observatory on Mauna Kea that does not have a telescope operator; the astronomer fully controls the telescope while observing.

A rainy day on Mauna Kea is bad news for astronomers on any telescope. However, even a sunny day can sometimes pose problems at submillimeter observatories. Water vapor in the atmosphere absorbs submillimeter light like a sponge absorbs water, preventing it from reaching the ground. To a submillimeter telescope, high humidity is like clouds: the moister the air, the dimmer the view of the heavens. Mauna Kea's usually very dry air is the reason the CSO was constructed there. However, weather systems that pass over the island occasionally bring clear but moist skies that make observations with submillimeter telescopes difficult or impossible.

The twin Keck telescopes at night, illuminated by the light of the full moon.

The revolutionary design of the Keck mirrors is reflected in the design of Keck headquarters.

Mauna Kea is home to the two largest optical and infrared telescopes in the world: the twin Keck telescopes. Built with a generous gift from the W. M. Keck Foundation, the Keck I Telescope saw first light in 1993, and its twin, Keck II, followed in 1996. Together they cost $140 million—less than 1 percent of the amount that Americans spend on dietary supplements each year. The sights they have revealed are priceless.

The sheer enormity of the Keck mirrors—10 meters (33 feet) in diameter—required a radical new design. Rather than being made from a single piece of glass, each Keck mirror is composed of thirty-six smaller mirror segments that fit together like pieces in a puzzle. Each individual segment is 1.8 meters (6 feet) across and weighs 880 pounds. Computer-controlled pistons underneath the segments keep them perfectly aligned with each other, to an accuracy of better than one-thousandth the width of a human hair. Keck's segmented design has proven so successful that the next generation of extremely large telescopes will likely be based on a similar design.

Although the two Keck telescopes are identical, each uses different instruments to explore the universe, and they usually work independently. However, it is possible to combine the light from both telescopes together to enhance their ability to see fine details. The W. M. Keck Observatory is managed by the California Association for Research in Astronomy, whose board of directors includes representatives from Caltech (California Institute of Technology), the various campuses of the University of California, and NASA.

Subaru Telescope

Japan's most powerful optical and infrared telescope is not in Japan—it is in Hawai'i. The Subaru Telescope takes its name from the Japanese word for the group of stars known as the Pleiades in English and Makali'i in Hawaiian. Mauna Kea's good weather, pristine air, and location near the equator (which provides access to the entire northern sky and part of the southern sky) were primary reasons for building this new-generation telescope four thousand miles from Japan.

Like those of most large telescopes, Subaru's birth was a long process—fifteen years from first planning to first light in 1999. The heart of the telescope is an ultrathin glass mirror 8 meters (26 feet) in diameter, which, although only 20 cm (8 inches) thick, still weighs twenty-three tons—as much as five full-grown elephants. A bevy of state-of-the-art instruments provides a wealth of information from the light collected from distant moons, planets, stars, and galaxies.

Great effort went into Subaru's design to achieve the sharpest views of the heavens. Subaru's protective dome is cylindrical rather than spherical, as are most other observatories' domes. This unusual shape was chosen after computer models and experiments suggested it would allow air to flow more smoothly around the telescope, enabling sharper images to be obtained without the distortions caused by air turbulence. Also, telescope operators and astronomers sit in an adjacent building rather than next to the telescope to help minimize heat within the telescope dome that could distort images. Future plans include enabling the telescope to be controlled remotely from Japan.

Canada-France-Hawai'i Telescope

The Canada-France-Hawai'i Telescope (CFHT) has been capturing stunning views of the heavens for over a quarter century. The observatory is a joint venture between the astronomical communities of Canada, France, and the University of Hawai'i. Completed in 1979 at a cost of thirty million dollars, its 3.6-meter (12-foot) mirror made it the sixth largest optical telescope in the world at the time. As one of the first observatories on Mauna Kea, CFHT had its choice of locations, and it is reputed to have perhaps the best site on the summit.

Instrumentation is often a great equalizer among telescopes. A cleverly designed instrument that collects and dissects the light from stars and galaxies in novel ways can make a smaller telescope every bit as valuable as a larger one. In the new era of giant 8- and 10-meter (26- and 33-foot) optical telescopes, CFHT has remained competitive by developing innovative instruments that provide astronomers with data that cannot be obtained elsewhere. In recent years, for example, the observatory has devoted much effort to wide-field imaging, building astronomical cameras capable of capturing the faint light from large swaths of sky.

The CFHT Legacy Survey currently under way to map large regions of the sky in unprecedented detail is providing astronomers with valuable data that will be mined for information for many years to come.

University of Hawai'i's 0.6-meter (24-inch) Telescope

The smallest and oldest telescope on the mountain, the University of Hawai'i's 0.6-meter (24-inch) Telescope was built by the U.S. Air Force in 1968: Its original mission was to track satellites; however, when it proved too slow to track many satellites, the air force donated it to the University of Hawai'i for astronomical research.

Despite its diminutive size and advanced age, this telescope has contributed to both research and education. In 1989 it became the first ground-based telescope to test newly developed infrared cameras that would later be used on the Hubble Space Telescope. Today it is used primarily as a learning tool by undergraduate astronomy students from the University of Hawai'i at Hilo. Observing here requires endurance, as it is the only observatory on the mountain with no heated control room and no toilet facilities. Plans are under way to replace the current telescope with a new, larger, 1.0-meter (39-inch) telescope that will be remotely controlled from Hilo.

The University of Hawai'i's 2.2-meter (88-inch) Telescope

The oldest telescope still in continuous nightly use on Mauna Kea, the University of Hawai'i's 2.2-meter (88-inch) Telescope, got off to a rather shaky start. Construction of this first sizeable telescope on Mauna Kea began in the fall of 1967, but two seasons of unusually harsh winter weather delayed its completion until 1970. Even then the problems were not yet over, as getting the complex telescope control system to work properly took several additional years of effort.

Despite these early problems, the telescope has proven to be an invaluable member of the community of telescopes atop Mauna Kea. A workhorse for the University of Hawai'i for over three decades, it has provided observing experience and data for faculty and students alike, including about one hundred PhD students' dissertations. Recent upgrades to the telescope allow it to be completely computer controlled from Hilo or Honolulu; only rarely are astronomers required to observe from the summit.

A second 0.6-meter (24-inch) telescope also once resided on Mauna Kea's summit. The University of Hawai'i's Planetary Patrol Telescope, built in 1970, was mainly used for observations of planets and other solar system objects. It was removed in 1995 to make room for the Gemini North Telescope that currently occupies the same site.

Very Long Baseline Array

The 240-ton dish of the VLBA on Mauna Kea towers seven stories above the ground.

The locations of all VLBA sites. Image courtesy of National Radio Astronomy Observatory/AUI/NSF.

If bigger is better when it comes to telescopes, then how about one the size of a continent?

Just below the summit of Mauna Kea, at an elevation of twelve thousand feet, sits a giant radio telescope. Built in 1992 and weighing 240 tons—the combined weight of nearly three thousand adults—it uses a massive dish 25 meters (82 feet) across to collect faint radio light emitted by distant stars and galaxies. It is one of ten similar telescopes, spreading from the Virgin Islands to Hawai'i, that together form the Very Long Baseline Array (VLBA). All ten VLBA dishes are controlled remotely from VLBA headquarters in New Mexico.

Contrary to popular belief, radio telescopes like the VLBA do not "listen" to signals from space; radio is just one of many forms of light collected by telescopes. (A car radio, for example, is essentially just a miniature radio telescope that collects radio light and converts it into an audible signal.) What astronomers get from radio telescopes like the VLBA are streams of numbers that measure the intensity of the radio waves striking them, not sounds. By pointing all ten VLBA dishes at the same object at the same time, it is possible to combine their light into one image. In this way the VLBA effectively acts like a single, continent-wide radio telescope that stretches over five thousand miles from end to end. This allows it to see much finer details than any other telescope on our planet. The VLBA's eyesight is so good that, in principle, it could read the time on a wristwatch four thousand miles away.

Submillimeter Array

Eight movable dishes make up the complete Submillimeter Array, seen here on a snowy day

The newest addition to the community of telescopes on Mauna Kea is the Submillimeter Array (SMA). It consists of eight separate dishes, each 6 meters (20 feet) in diameter, that are linked together to form one giant submillimeter telescope. Built as a partnership between the Smithsonian Astrophysical Observatory at Harvard and the Institute of Astronomy and Astrophysics in Taiwan, construction of the SMA began in 1995 and was completed in 2003. Unlike most telescopes, which do not budge, each dish of the SMA can be moved to any of twenty-four different locations on the summit. The advantage of a movable array of dishes is that it allows the telescope to act like a giant zoom lens. When the eight dishes are placed at their widest separation, the SMA's vision is as sharp as that of a single enormous telescope 500 meters (1,640 feet) in diameter—about the size of five football fields. This provides amazingly sharp views of small regions of the sky, revealing far more details than any stationary single-dish submillimeter telescope could. Moving the dishes closer together, on the other hand, acts like a wide-angle lens to capture images from a larger region of the sky, but with less detail. Like other submillimeter telescopes, the SMA is an especially powerful tool for studying the cold, dusty regions of the universe where stars and planets are being born today.

The Astronomical Village

RIGHT: A pioneer of astronomy in Hawai'i, Walter Steiger established Hawai'i's first modern astronomical observatory at Makapu'u Point on O'ahu in 1957, and another in 1961 on Maui. Steiger's success encouraged others—most notably Gerard Kuiper—to build astronomical observatories in Hawai'i. "Mauna Kea is truly a very special place," says Steiger. "At the summit we can drink in the wisdom and beauty of the gods: the heavens full of stars, the Milky Way and myriad worlds of mystery. There lie the answers to our questions about the origins of man and man's place in the universe. And yet Mauna Kea is more than the sum of all its parts. It is a place that demands our awe and reverence. At the same time, it is a fragile environment that can be smothered and damaged by too much admiration. We must learn to admire and respect this unique spot of Earth with the least possible impact. And we must respect the rights of all our fellow beings to share in the glory of this mountain. These things I believe." Steiger poses here in front of a photo of the sun's roiling surface taken in 1958 with the telescope at the Makapu'u Point Observatory.

OPPOSITE PAGE: As the sun sets, an astronomer scans the sky from the catwalk of the Canada-France-Hawai'i Telescope.

Lawe i ka ma'alea a kū'ono'ono.
Take wisdom and make it deep.
~ '*Ōlelo No'eau*, Mary Kawena Pukui, collector and translator, Bishop Museum Press

When the ancient Hawaiians prepared to embark on their epic voyages across the sea, it took a community-wide effort. Although only a few chosen individuals set sail for distant lands, everyone pitched in to build the voyaging canoes, weave the sails, and prepare food and supplies for the long journeys. Without the support and sacrifice of so many people, the voyagers

might never have reached their destinations. It is the same with astronomy on Mauna Kea. The nightly "voyages" of scientific discovery there depend on the efforts of many people.

The astronomers

Human beings have always watched the skies, so it is not surprising that astronomy is the oldest of all sciences, with roots that go back to the dawn of civilization. Today, astronomers on Mauna Kea continue the proud tradition of stargazing begun by our ancestors so long ago. Although the technology has changed over time—from naked-eye astronomy to multimillion-dollar telescopes—the quest remains the same: to understand our origin, our place in the cosmos, and perhaps our destiny. One of modern astronomy's most profound discoveries is that we humans are made from the ashes of stars whose fires burnt out billions of years ago, our atoms forged by nuclear fusion deep in their interiors and then strewn throughout space by their explosive deaths as supernovas. We are truly star stuff. Perhaps this is why we feel compelled to explore the starry skies, as if driven by an innate yearning to know our true ancestral home.

Astronomers come from around the world to use the telescopes on Mauna Kea. While there they sleep, eat, work, and socialize in the astronomers' residence known as Hale Pōhaku, or "Stone House" in Hawaiian, at an elevation of 9,300 feet. To avoid the dangers of altitude sickness, nobody is allowed to sleep at the summit. In fact, the observatories adhere to a strict "two person" rule, meaning that no one is ever alone at the summit—a precaution should illness strike.

There are about ten thousand professional astronomers around the world. Most work at universities and colleges, where their duties include both teaching and research. Some work in government agencies such as NASA or at observatories. Hundreds of astronomers from all over the world come to Mauna Kea each year to use the telescopes. Many more wish they could. The demand for viewing time on Mauna Kea telescopes far exceeds the number of available nights, so astronomers must compete for telescope time by submitting requests as much as a year in advance.

A panel of experts from each telescope then faces the difficult task of deciding which of the many excellent scientific project proposals are most likely to advance the field of astronomy. Competition is fierce and an astronomer is fortunate to be awarded a few nights of observing time on Mauna Kea telescopes each year. Astronomical research requires great patience because it can often take years to accumulate enough data to complete a project. Time with the telescopes on Mauna Kea is so precious that even when skies are overcast, astronomers stay up all night hoping for a brief break in the clouds. Sometimes even just a few minutes of clear skies are all that's needed to achieve the thrill of discovering something never before seen.

A typical observing run lasts anywhere from one night to one week. Thanks to computers, astronomers no longer need to sit huddled inside a cold observatory dome all night. Instead,

ABOVE: The astronomers' dorms at night.

LEFT: A meteorologist provides daily weather forecasts for the summit that include predictions for nighttime-viewing conditions. On average nearly three hundred nights a year are clear on Mauna Kea, and a beautiful canopy of stars blazes overhead. But all the high-tech equipment in the world can't change the weather, and some nights the skies refuse to give up their secrets, leaving disappointed astronomers to wonder what discoveries might have been.

RIGHT: Astronomer Michael Gregg of the University of California at Davis uses the Keck II Telescope to collect data on a newly discovered class of galaxies known as "ultracompact dwarfs." While Gregg controls the telescope's camera remotely from Keck headquarters in the town of Waimea, the telescope operator seen on the television screen behind him controls the telescope from Mauna Kea's summit. Gregg, a frequent visitor to the mountain, says, "Astronomy is usually considered to have no practical value. But when I look up at the summit of Mauna Kea and see the international huddle of giant telescopes, I'm reminded that pure science can inspire humans to rise to our best in pursuit of the answers to age-old questions that seemed unfathomable just a generation ago. In the present world, what could be more valuable than that?"

RIGHT: University of Hawai'i astronomer Marianne Takamiya (right) and student Kumino Miura (left) observing distant galaxies from the control room of the Subaru Telescope on the summit of Mauna Kea.

OPPOSITE PAGE: As dusk settles over Mauna Kea, support scientist Tony Matulonis checks the Gemini North Telescope in preparation for a night of observing. Observing assistants such as Matulonis operate the telescope for the astronomers, pointing it at the desired locations in the sky, focusing it, and troubleshooting when problems arise.

Kimo Pihana is a ranger on Mauna Kea. A respected elder in the Native Hawaiian community, Pihana works to protect this mountain that he holds so sacred, and all who visit it. "I really enjoy working up here because of my commitment and dedication to my culture and my community," he says. "It has made me dwell on some of the past, walking in the footprints of our ancestors." A ranger's day on Mauna Kea is a long one that includes monitoring road conditions, providing medical assistance in emergencies, and answering questions from visitors. "We're here to protect not only this mountain but the whole island," says Pihana. "I've got so much to share."

OPPOSITE PAGE: Despite the hardships of high altitude, cold temperatures, and thin air, the summit of Mauna Kea has become an increasingly popular spot for tourists and Hawai'i residents alike. Many make the journey to see the astronomical observatories, enjoy the spectacular sunsets, and stargaze.

they sit in a warm, comfortable control room adjacent to the telescope, with telescope assistants and modern conveniences such as stereos and microwave ovens. Nevertheless, working above the clouds is very demanding, both physically and mentally. The altitude at the summit of Mauna Kea is fourteen thousand feet—about half as high as a jet plane flies—and there is 40 percent less oxygen than at sea level. Here, at the edge of Earth's atmosphere, it is easy to become fatigued and forgetful as the body struggles with oxygen deprivation and dehydration in the rarefied air. Altitude sickness is common, with symptoms ranging from mild headaches, nausea, shortness of breath, and dizziness to (occasionally) more life-threatening conditions. For many astronomers the discomforts of high altitude are compounded by jet lag after a long flight to Hawai'i.

The Visitor Information Station on Mauna Kea, located midway up the mountain, welcomes more than one hundred thousand visitors a year. Staffed mostly by volunteers, it offers lectures on astronomy and Hawaiian culture, public stargazing with telescopes, and hot chocolate to help visitors stay warm in the cold night air.

OPPOSITE PAGE: People come to Mauna Kea for many reasons. Some come to study the cosmos with powerful telescopes. Some come to worship on the sacred land of their ancestors. Some come to explore the mountain's remarkably diverse landscape and unique ecosystems. Some come to play in the winter snow.

Perhaps in the not-too-distant future, astronomers will no longer need to travel to the summit of Mauna Kea to use the telescopes. To combat the hardships of high altitude, more and more observatories have begun to offer remote observing, which allows astronomers to control the telescope and collect data via the Internet from many miles away. The twin Keck telescopes, for example, now operate almost exclusively in remote-observing mode, with astronomers working from control rooms at Keck headquarters an hour's drive away in the scenic town of Waimea. A photograph on the wall above the astronomer's computer shows the Keck telescopes glistening in the light of the setting sun, with an arrow that reads, "You are observing here."

Once the observing run is completed and the astronomers return home, they usually spend weeks or months analyzing the data, trying to coax every bit of available information from the light collected by the telescope. Astronomers generally spend most of their time working on computers, analyzing data, teaching university classes, writing applications for more telescope time, writing scientific papers summarizing their research, reading papers written by

other astronomers, and presenting their discoveries at international astronomy conferences. It is hard work, but it is also tremendously exciting and rewarding. Most astronomers believe they have the best job on the planet.

A hard day's night

When the first light of dawn peeks over the horizon, astronomers on the summit of Mauna Kea, tired after a long night of observing, close their telescope domes and drive down the mountain for some well-deserved sleep at Hale Pōhaku, or "Stone House," the astronomers' residence. For others on Mauna Kea, however, the workday is just beginning.

Observatories are around-the-clock operations. For every astronomer on Mauna Kea, five to ten support people work on the summit or in offices below. Engineers build and test new instruments for the telescopes. Technicians repair or replace worn parts to keep everything functioning properly. Administrators make reservations and arrange transportation for a constant stream of astronomers traveling to and from the mountain. Computer programmers write software to control the telescopes and archive the data they collect. Cooks prepare meals for hungry astronomers and observatory staff at Hale Pōhaku. Road crews clear snow, ice, and rocks. Without the support of this diverse community of dedicated professionals, none of the astronomical discoveries made on Mauna Kea would be possible.

Tremendous effort goes into keeping the telescopes operating in tip-top condition. The stakes are enormous. At an estimated cost of as much as fifty thousand dollars per night— nearly one dollar *per second*—for observing time on Mauna Kea, the last thing anybody wants is to lose precious moments during the night because of mechanical problems with the telescope or its instruments. So while astronomers sleep during the day, technicians work to ensure that these complex machines are ready for the next night of observing. It is challenging work; to see the faintest and farthest objects in the universe with 20/20 vision requires pushing the limits of technology. Astronomical cameras must be chilled to temperatures near absolute zero (-459 degrees Fahrenheit) to increase their light-gathering power. Telescopes weighing hundreds of tons must be balanced so perfectly that they can be moved with the precision of a fraction of the width of a human hair. Cameras bigger than a refrigerator must be swapped on and off the telescopes depending on the needs of the observations planned for the coming night. Moreover, this all must be done at an altitude of fourteen thousand feet, where even thinking clearly can be a challenge. However, it is done every day by dedicated support crews.

OPPOSITE PAGE: There are places in the universe that even the Mauna Kea telescopes cannot show us. These are left for the imagination—or the hands of a skilled artist such as Jon Lomberg, the "man who paints the stars." One of the world's best-known astronomy-inspired artists, Lomberg's many honors include an Emmy Award for his work as chief artist for Carl Sagan's groundbreaking *Cosmos* television series, and an asteroid that was named after him. His art is literally out of this world; he designed the cover of the Voyager interstellar record, a gold-plated phonographic record carrying images and sounds of Earth that was launched with the Voyager 1 and 2 spacecrafts on their journeys beyond our solar system, destined to wander the Milky Way perhaps forever. This Big Island resident has also illustrated scientific discoveries made by telescopes on Mauna Kea, helping to increase public awareness and understanding of astronomical research. "If you're interested in astronomy, now is absolutely the best time to be alive and working," says Lomberg. "All these many discoveries need to be explained, not only with words but with pictures. When the discoverer of something likes my representation of it, it makes me feel almost like I'm part of the discovery team, at the cutting edge of science. It's very exciting and thrilling." Artwork courtesy of Jon Lomberg.

Kahalelaukoa Ka'ahanui EII, known simply as "Koa," is a cultural guide on Mauna Kea, sharing her *aloha* and her knowledge of Hawaiian culture with visitors. Her name reflects her rich heritage: Kahalelaukoa refers to the "house of four hundred women warriors," and Ka'ahanui means "great gathering." "We Hawaiians have always been *kilo hōkū*," she says. "We have always been stargazers. The *kūpuna* [elders] said that we are particles of the heavens; the scientists today say that we are particles of the stars. So we need to look at the commonalities between science and culture, not at the differences. I'm grounded here on Mauna Kea. The *mana* of my ancestors, the spiritual power of my ancestors, is here. I can feel them."

OPPOSITE PAGE: Mauna Kea seen from the neighboring mountain of Mauna Loa.

Mauna Kea's diverse community also extends beyond the astronomical village. Rangers patrol the mountain to protect people and places. Cultural interpreters share the living culture of the Hawaiian people and their special relationship with this sacred mountain. Tour guides bring vans of curious visitors to see the telescopes and watch the sunsets. Local residents pile freshly fallen mountaintop snow into the backs of pickup trucks to build snowmen on the beach. Student volunteers set up telescopes for public stargazing at lower elevations. The spirit— *mana*—of this special place touches all who come to Mauna Kea.

According to a Hawaiian proverb, *E lauhoe mai na wa'a, pae aku i ka 'āina*, "Everyone paddles the canoe together; and the shore is reached." (*'Ōlelo No'eau*, Mary Kawena Pukui, collector and translator, Bishop Museum Press.) On a quiet night on Mauna Kea, as the telescopes peer into the starry skies, one can almost hear the faint sound of hundreds of oars being dipped into the waters of the cosmic ocean.

Voyages of Discovery

Image by Bryan Lowry.

RIGHT: Sunsets on Mauna Kea often produce a dazzling display of colors.

OPPOSITE PAGE: Shining like brilliant blue diamonds against the velvety blackness of space, this cluster of blue stars is known as M35. A second, more distant cluster of yellowish stars called NGC 2158 can also be seen. Most stars, including our own sun, may have been born in stellar groups like these that slowly disperse over time, scattering their stars throughout the galaxy. The difference in color between these two stellar systems reveals differences in their ages; the stars in NGC 2158 are ten times older than those in M35. This photo was obtained by the Canada-France-Hawai'i Telescope; total exposure time was fifteen minutes. Image courtesy of Canada-France-Hawai'i Telescope.

Mai ka pō mai ka ʻoiāʻiʻo.
Truth comes from the night.
~ *'Ōlelo No'eau*, Mary Kawena Pukui, collector and translator, Bishop Museum Press

Tonight, as glittering celestial bodies begin to appear in the sky above Mauna Kea, a new journey into the unknown begins. Astronomers make their final preparations. Observatory domes awaken from their daytime slumber and slowly creak open. Finally, after the last glow of twilight has faded, telescopes peer intently into the starry skies overhead until the morning light returns.

More than fourteen thousand nights have passed in this way since the first telescope arrived on this remote mountaintop in the middle of the Pacific Ocean. In that time,

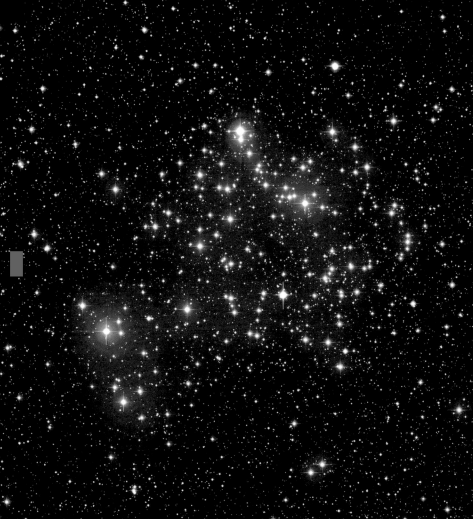

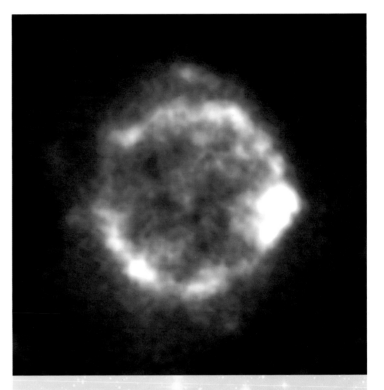

Mars, the red planet, has fascinated people throughout history. It is usually considered the most likely planet besides Earth to have life, if not today then in the past, when the planet was undoubtedly warmer and wetter. This infrared image of Mars was taken with the United Kingdom Infrared Telescope in August 2003, while the planet was at its closest approach to Earth in nearly sixty thousand years. Made from a series of very short exposures, each less than a fraction of a second, it is perhaps the sharpest image of Mars ever obtained with a ground-based telescope. The appearance of dark and light regions is caused by differences in the composition of surface rock exposed by high winds that sandblast the Martian surface. The planet's south polar ice cap is visible at the bottom. Image courtesy of Jeremy Bailey and the Joint Astronomy Centre.

The death of a star. The James Clerk Maxwell Telescope captured this image of Cassiopeia A, a star that exploded as a supernova about ten thousand years ago. Although such violent stellar explosions are rare—occurring only about once a century in galaxies like ours—they are important because they enrich space with stardust, the raw material from which new generations of stars, planets, and perhaps life on other worlds will be born. This ghostly image shows the submillimeter light emitted by Cassiopeia A during its last gasps, as its outer layers expand into space, propelled by the force of the explosion that occurred deep within the star's interior. Image courtesy of Loretta Dunne and the Joint Astronomy Centre.

"Silently one by one, in the infinite meadows of heaven, blossomed the lovely stars, the forget-me-nots of the angels." Henry Wadsworth Longfellow (1807–1882). The photo here, taken by the author using the University of Hawaiʻi's 2.2-meter (88-inch) Telescope, shows a cluster of stars known as M71. The many thousands of stars that make up this cluster are all about fourteen billion years old, making this one of the oldest objects in the universe. First discovered by a Swiss astronomer in 1745, it is one of about 150 similar clusters that are now known to exist throughout our Milky Way galaxy.

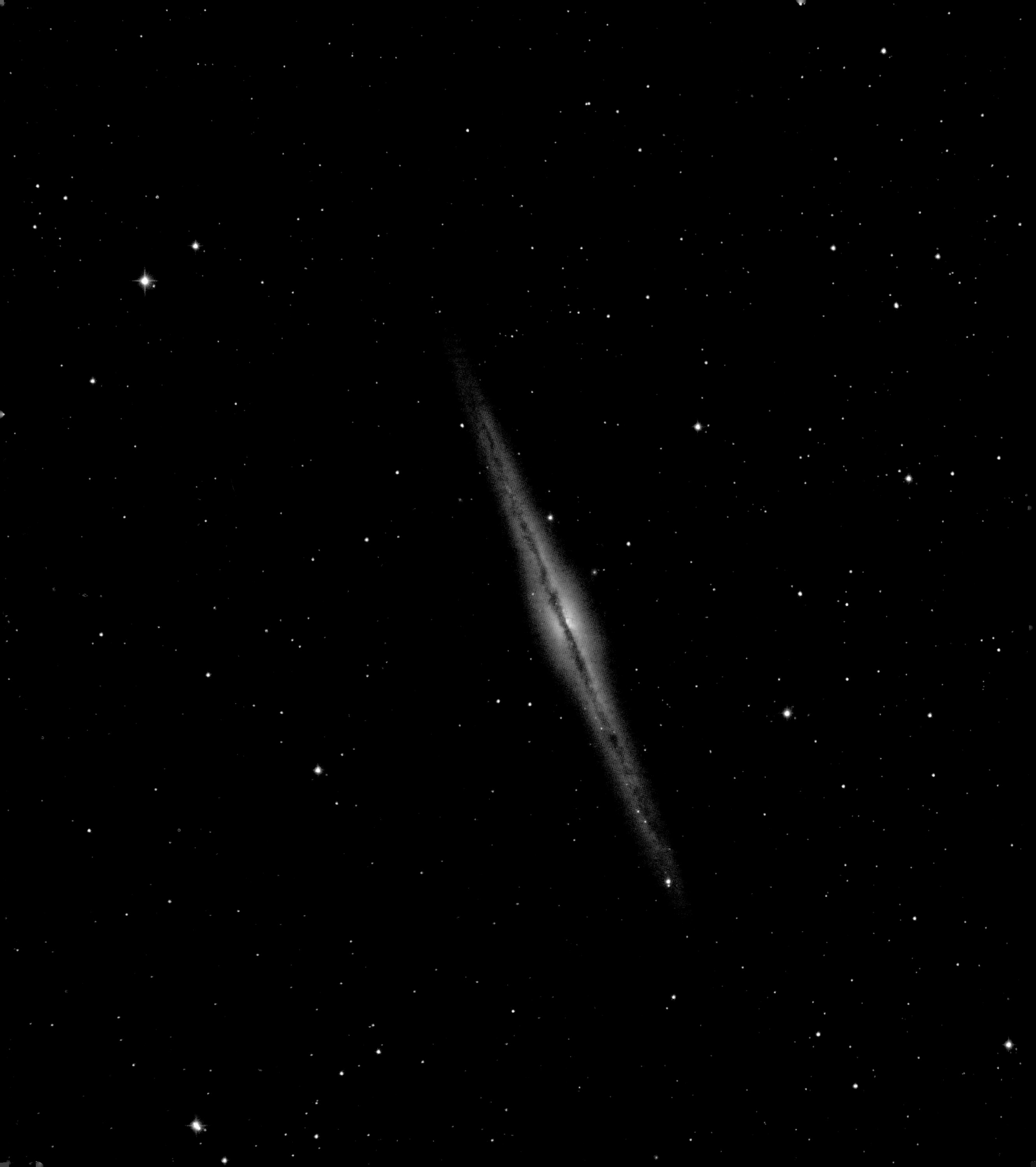

Saturn is not the only planet in our solar system with rings. The sequence of infrared photos here, taken by NASA's Infrared Telescope Facility, shows Jupiter's rings, as well as two of its moons, Metis and Amalthea, as they orbit nearby. Image courtesy of John Rayner and NASA IRTF.

Mauna Kea's observatories have been at the forefront of astronomical exploration. Each telescope, from the largest to the smallest, has contributed to advancing our knowledge of the universe.

The list of discoveries made with the telescopes on Mauna Kea is long and impressive. Subaru has captured the faint light of infant galaxies at the very edge of the observable universe, providing important clues about the birth of galaxies. The Canada-France-Hawaiʻi Telescope (CFHT) has found dozens of previously unknown

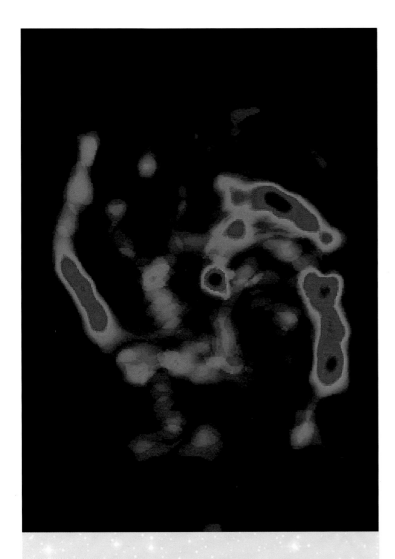

A submillimeter portrait of a galaxy known as M51, taken by the Submillimeter Array. A spiral pattern can be seen, similar to the galaxy in the image on page 99. Whereas the visible light in the photo on page 99 comes mostly from stars, the submillimeter light here comes from cold gas sprinkled throughout this galaxy. Image courtesy of K. Sakamoto and the SMA Extragalactic Team.

Things that go bump in the night. The image here, taken by the Caltech Submillimeter Observatory, shows submillimeter light coming from two galaxies undergoing a violent collision. Unlike collisions between cars, which are over in an instant, collisions between galaxies take place over millions or billions of years, timescales that are still only the blink of an eye compared to the lifetime of a typical galaxy. Galaxy cannibalism is also common, as larger ones devour smaller ones, inheriting their stars, planets, and other material in the process. Or a gust of gravity from a passing galaxy can rip stars loose, creating an ever-growing sea of orphaned stars. Locked in a gravitational embrace that neither can escape, the two galaxies shown here will eventually merge into a single, larger one billions of years from now. Image courtesy of the Caltech Submillimeter Observatory.

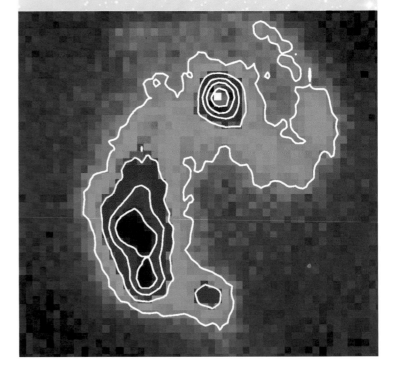

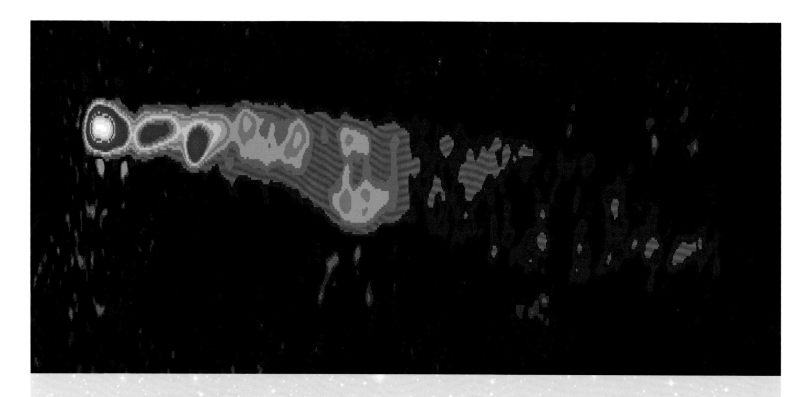

Lurking in the centers of many galaxies are black holes of unimaginable size—millions or even billions of times more massive than our sun. Black holes are gravity's ultimate triumph, a place where the gravitational pull is so strong that nothing—not even light—can escape. The image here, taken by combining data from the ten telescopes that make up the Very Long Baseline Array, shows an intense jet of radio emission being ejected near a supermassive black hole as it devours material that strayed too close and fell into its gravitational grip. Image courtesy of R. C. Walker and NRAO/AUI.

moons in our solar system, and clues to how they might have formed. The Keck telescopes have revealed more planets orbiting other stars than any other telescopes, confirming centuries of speculation that other solar systems like ours are common. Gemini North has glimpsed enormous groups of thousands of galaxies all moving in a gravitational dance around each other. The James Clerk Maxwell Telescope (JCMT) has mapped the cold material that lies between the stars, revealing in finer detail than ever before where the next generation of stars will be born long after our sun is gone. The University of Hawaiʻi's 0.6-meter (24-inch) Telescope has found

Like snowflakes in a cosmic blizzard, galaxies are strewn throughout the universe. The Subaru Telescope spent more than two hundred hours collecting light from one particular region of space to make this spectacular image. The light of more than a million galaxies is contained in this image, which covers an area of sky equivalent to the width of a finger held at arm's length. The faintest galaxies in this image are over one hundred million times fainter than the naked eye can see. An image like this allows astronomers to study enormous numbers of galaxies and to map their distribution in space. It is also a slice through time, since it reveals galaxies at different distances from Earth, and hence of different ages. Image courtesy of Subaru Telescope, National Astronomical Observatory of Japan.

The space between stars is almost, but not completely, empty. Thin wisps of gas and stardust are found throughout our Milky Way galaxy, occasionally condensing into billowy clouds. These regions are where the next generation of stars will be born. The image here, taken with the James Clerk Maxwell Telescope, shows the glowing emission of gas and dust toward the center of our galaxy. It is the largest, most detailed map of the submillimeter light from our galaxy's center that has ever been obtained. Image courtesy of Douglas Pierce-Price and the Joint Astronomy Centre.

evidence of "star quakes" on nearby stars. NASA's Infrared Telescope Facility (IRTF) has measured wind speeds in the atmosphere of Saturn's moon Titan from almost a billion miles away. Yet, just as important as the telescopes and their cutting-edge instruments are the men and women who use them. The many exciting astronomical discoveries made on Mauna Kea are more than just feats of technological prowess; they are triumphs of human ingenuity and the spirit of exploration.

Nobody can predict what secrets of the cosmos the darkness may reveal tonight. We are living in a golden age of astronomical exploration, with astonishing discoveries being made almost daily. Knowledge gained by astronomers on Mauna Kea

If we could see our home galaxy, the Milky Way, from far off in space, it would probably resemble this beautiful spiral galaxy, known as NGC 628. Nestled within its pinwheel pattern are massive star clusters, gas clouds, and regions of dark dust. This galaxy is located about thirty million light-years from Earth, which means that we see it as it appeared thirty million years ago, when the light that made this image left on its journey to Earth. This photo, obtained with the Gemini North Telescope atop Mauna Kea, was made by combining three separate photos taken with blue, green, and red filters. Spiral galaxies like this are common throughout the universe. Image courtesy of Gemini Observatory and the GMOS Team.

The history of the universe is essentially a tale of atoms being continuously rearranged in new ways. The infrared image here, taken with the Subaru Telescope, shows a stellar "nursery," where stars are being born from gas and dust recycled from previous generations of stars. As they begin to shine, the light from these energetic newborn stars causes surrounding gas to glow like cosmic neon lights. On average, only about one new star forms each year in our Milky Way galaxy, though astronomers believe the rate must have been much higher in the past, since at the current laid-back pace of stellar birth, it would have been impossible to form all of the hundreds of billions of stars that make up our galaxy within the 13-billion-year history of the universe. Indeed, tremendous bursts of star formation are sometimes seen in other galaxies, suggesting that our galaxy once had a more energetic youth. The sun, Earth, and all the planets of our solar system were likely born in a similar gas cloud about 4.6 billion years ago. Image courtesy of Subaru Telescope, National Astronomical Observatory of Japan.

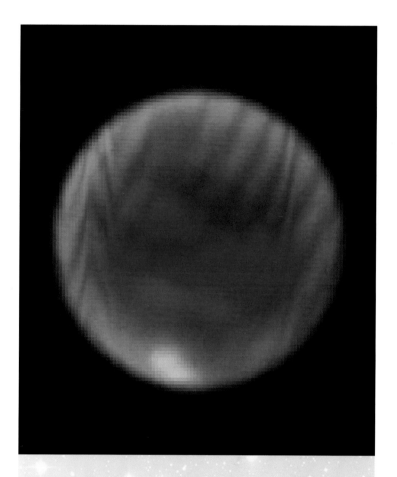

Saturn's moon Titan is the only moon in our solar system that possesses an atmosphere, a discovery made by Gerard Kuiper in 1944. Astronomers consider Titan, like Mars, to be a place where life might exist, or might once have existed. Infrared imaging by the Keck telescopes allows astronomers to monitor changing cloud patterns on this intriguing moon. The bright region at the bottom is the largest storm ever seen in Titan's atmosphere. Image courtesy of W. M. Keck Observatory/ Antonin Bouchez.

ABOVE: Holes in the clouds allow the Canada-France-Hawai'i Telescope to glimpse the stars overhead.

OPPOSITE PAGE: Today's children are tomorrow's explorers on Mauna Kea.

tonight will be passed on for the generations to come, just as it was by those who came before them. But we human beings are newcomers to the cosmos, having existed for less than 0.1 percent of Earth's 4.6-billion-year history, a brief instant in the 14-billion-year timescale of our universe. We still have so much to learn. The universe is filled with wonders not yet seen or even imagined that await our discovery. It is the thrill of voyaging into the unknown and returning with perhaps the greatest treasure of all—knowledge—that keeps astronomers returning to Mauna Kea again and again.

Appendix

Additional information about astronomy on Mauna Kea can be found on the following Web sites:

Observatories

Caltech Submillimeter Observatory
www.submm.caltech.edu/cso/

Canada-France-Hawai'i Telescope
www.cfht.hawaii.edu/

Gemini Observatory
www.gemini.edu/

James Clerk Maxwell Telescope
www.jach.hawaii.edu/JCMT/

Keck Observatory
www2.keck.hawaii.edu/

NASA's Infrared Telescope Facility
http://irtfweb.ifa.hawaii.edu/

Subaru Telescope
www.subarutelescope.org/

Submillimeter Array
http://sma-www.harvard.edu/

United Kingdom Infrared Telescope
www.jach.hawaii.edu/UKIRT/

University of Hawai'i's 24-inch and 88-inch telescopes
www.ifa.hawaii.edu/88inch/

Very Long Baseline Array
www.vlba.nrao.edu/

Others

Mauna Kea Astronomy Education Center
www.maunakea.hawaii.edu/

Mauna Kea web cameras
http://mkwc.ifa.hawaii.edu/current/cams/

Onizuka Center for International Astronomy, Mauna Kea Visitor Information Station
www.ifa.hawaii.edu/info/vis/

University of Hawai'i at Hilo Department of Physics and Astronomy
www.astro.uhh.hawaii.edu/

University of Hawai'i's Institute for Astronomy
www.ifa.hawaii.edu/

OPPOSITE PAGE: The Very Long Baseline Array (VLBA) at sunset.

About the Author

Michael West is a professor of astronomy at the University of Hawai'i. He received his PhD in astronomy from Yale University in 1987, and has held research and teaching positions in the United States, Canada, the Netherlands, and the Gambia (West Africa). For over a decade, he has used the telescopes on Mauna Kea to study the formation and evolution of galaxies, and the structure of the universe on the largest scales. He lives in Hilo with his wife and son.

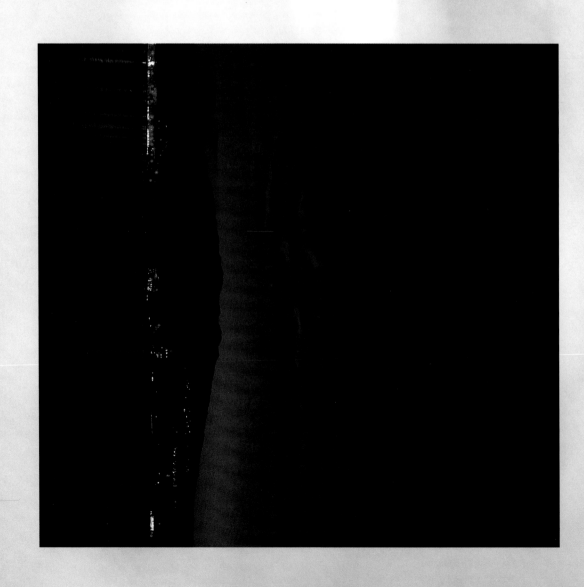